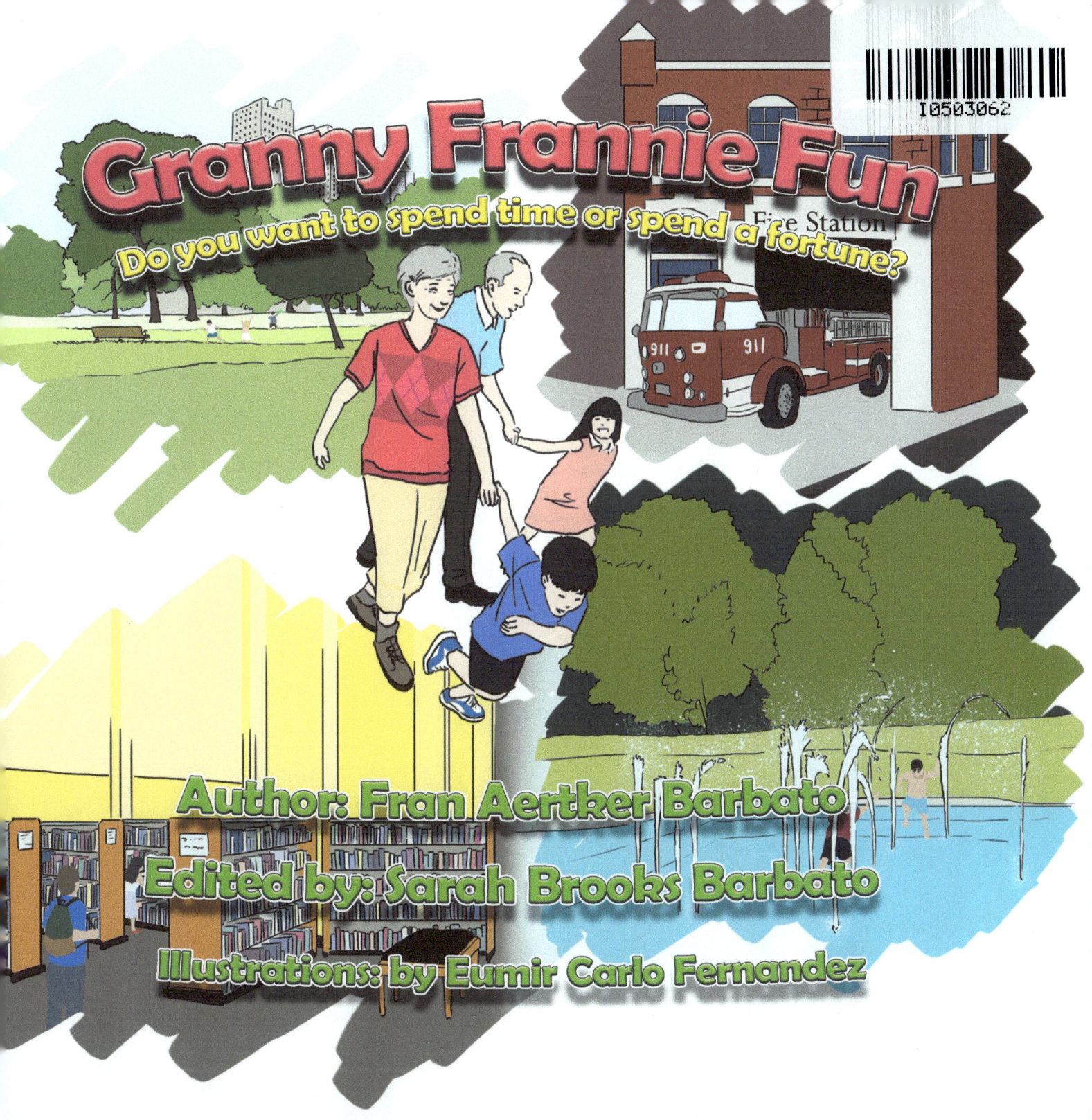

Granny Frannie Fun

Do you want to spend time or spend a fortune?

Author: Fran Aertker Barbato

Edited by: Sarah Brooks Barbato

Illustrations: by Eumir Carlo Fernandez

To order additional copies of this book, contact:
Xlibris Corporation
1-888-795-4274
www.Xlibris.com
Orders@Xlibris.com

CONTENTS

Dear Grandparents,

While enjoying your role as a grandparent, your little one will soon crawl, talk, and walk. Before you know it, your grandchild will be at your back door with a suitcase in hand. Now, what are you going to do for four days together?
That's why I wrote this book!

As grandparents we have the unique opportunity to teach our grandchildren in ways that parents may not have the time. I was inspired to invest more time into my grandchildren by a verse from the Bible, Psalm 145: *"One generation will commend your works to another; they will tell of your mighty acts."*

As a lifelong educator, teaching was my natural instinct when having the opportunity to spend a week with my 3-year old grandsons. (This was to give their parents a "real" vacation.) I began thinking of the fun activities we would do at *Frannie's Camp*. Keeping their developmental levels in mind, my journey began. As the children grew and siblings came along, the ideas grew as well.

I soon realized my friends, who were becoming grandparents, were also looking for meaningful things to do with their grandchildren. So, this book is an assortment of fun activities that I've personally tested while teaching school or teaching my own grandchildren. Activities are divided into educational sections with age recommendations*.

Read through the book, and find what peeks your interest (because we know if you are interested, your grandchild will be too.) Don't try to do something that you are not comfortable doing. Do things that are fun for you, so you can laugh and enjoy this special time with your grandchild.

May both you and your grandchild create memories to last a lifetime! And may God bless your time with them.

Frannie

*Note: Age recommendations are noted by shapes: the triangles for 3-year olds; the parallelogram for 4-year olds; and the pentagon for 5-year olds, etc.

Science Fun

Titanic Challenge

Will it float or sink? Fill a large plastic tub or small wading pool with water to set up the activity. (Collect items before the grandchildren arrive. Begin to store items in your granny box.) Which will float, and which will sink?
Tried and true items that work: pine cone, rock, pencil, golf tee, plastic milk lid, penny, old keys, and clothespin

Calling all scientists (no lab coat needed)

No lab coat needed to have fun with surface tension. Give child a cup of water, an eyedropper, and a penny. Count how many drops of water will drop onto the penny before the water spills over. Guess what? You are now teaching surface tension. Try showing this on a leaf. If it rains, your little one can find leaves with bubbles on them.

Up, Up and Away

Ready to make your own bubble fun?   Combine 1 cup of Dawn™ liquid with about 2 T. karo syrup in a large plastic bin (or small swimming pool) filled with water. Begin to dip various items into the bubble solution to see which ones make large bubbles. The plastic surrounding a 6-pack of canned drinks makes great circles. Find other plastic circles such as funnels, rings, anything that is round. Make your own bubble maker by tying a string on a stick at the top and then further down the stick making a large loop. Then dip it into the bubbles. Then ask, "Are all bubbles going to be round?". Try it out using a square object. You can also use a wire coat hanger to form other shapes.
Great time for a photo op!

Ready to Blow!

To make a volcano, you will need an empty corked bottle, white vinegar, and baking soda. First put about 2-3 tablespoons of baking soda into the empty bottle, then add 1/2 cup of vinegar—quickly add the cork to the top and move away from the bottle. Get ready for the chemical reaction and watch it "blow" the cork out of the bottle! This is an **outside** activity. You can add food coloring to the vinegar for visual effect.

Let there be light!

Build a simple circuit using a large flashlight battery, two pieces of plastic-coated wire and small flashlight bulb. Connect one end of wire A (pull back some of the coated wire) to the positive side of the battery. Connect the other end of wire A to the flashlight bulb. Connect wire B to the negative side of the battery and also to the flashlight bulb…. what fun it is to see when the light comes if you have made a complete circuit. For those who are ready to experiment, try adding a switch or a bell to the battery and wires.

Stuck like glue.

Magnets are always fun for children. Start with one strong magnet (these usually cost a little more). Hold it under a table and see what magic you can move on top of the table with paper clips or nails. Give children other objects (paper clips, money, keys) that attract and repel. Now you have taught two good vocabulary words.
You can also teach that magnets have a North Pole and South Pole using a bar magnet. Put a needle on a small piece of Styrofoam (just larger than a needle) in a small bowl of water. Now hold magnet above the needle, and watch it spin until the needle points north.

Going on a bear hunt

Maybe we won't hunt bears in your neighborhood, but a scavenger hunt is fun for all ages. This is a great way for grandparents to walk the neighborhood with your grandchild. Grandparents may see things they have never seen before!
There are several ways to do it. You can make a list of items you know are in your

neighborhood or you can cut out pictures of items and put the pictures onto an empty egg carton. Suggestions: twig, yellow flower, bird's feather, one piece of trash, a rock, a leaf with branched veins, a leaf with parallel veins, a rock, a pine cone, an acorn, a pecan, 3 leaves: small, medium, large.

Is it a bird? A plane? No, it's
A kite! Don't try to make one … just purchase one. It's more fun to get outside and fly it! And don't forget to add a few paper airplanes…show them how to fold paper into making airplanes. Don't forget kite string.

Abracadabra!
Discover a little magic before your eyes when
ping pong balls roll to you and away from you with just a comb, and a piece of wool (use a sweater or skirt). Rub the comb like crazy on the wool piece to give it a charge, then set the ping pong ball on a flat table, and bring the comb close to it. Reason it moves? The comb has negative charges on it, and the ping-pong ball has no charge. The electrons in the comb repel the electrons in the ball, exposing the positive charges. It's magic!

Monet in Training

Using a paper coffee filter and food coloring, allow your grandchild to experiment mixing drops of food color onto the coffee filter. You will have beautiful works of art hanging in no time! Hang the filters with a clothespin outside, and watch to see which colors run the fastest!

Look at me!

Draw the outline of the child's body on butcher paper or any large roll of paper, and then give the child markers to fill in eyes, ears, and heart! Teach a little physiology. They are always surprised to see how big they are when looking at the flat picture. If your background is in the medical profession, this may be a good time to show where the heart, kidneys, spleen and stomach are located.

Inching along...

Worm races—if you are a gardener, this is quite fun. Dig up some earthworms and line up on your driveway. Give your child a spray bottle with water. Draw a chalk line to start the race. Spray the worms (they will begin to wiggle) to see which worm wins the race! Then return the worms to the garden, so they can continue their job of turning the dirt.

Music Note Fun

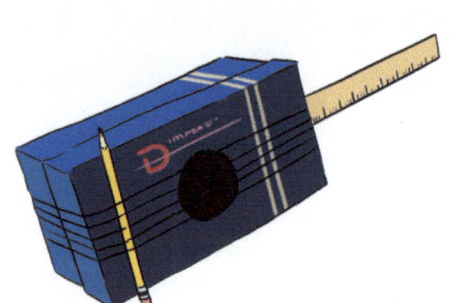

Music Note Fun

Rock On!

If you have never heard children make their own music, you are in for a treat. Make your own band...

Using an empty shoebox (no lid) or tissue box, stretch rubber bands of different thicknesses across the length of the box. Strum the bands and you will hear different tunes. For extra appeal, glue or tape a 12-inch wooden ruler to the bottom of the box simulating a guitar. Now it's time to decorate your guitar with paint and jewels. (Local craft store)

Need more instruments in your band?

- Create a drum using an empty cardboard oatmeal canister.

- Spice things up with a tambourine. Easy as pie to make... staple two paper plates together filling with either rice, red beans, or pennies to make different disposable tin pie plates work well as tambourines, also.

- Locate 5-6 glass bottles of differing sizes and shapes. Then fill each with a different amount of water. Add a different food coloring to each one. Gently blow across the top of the bottle to produce a musical note. You now have an organ.

- Find two pieces of wood about 4 inches long and 2 inches thick. Wrap with sandpaper for another instrument.

- Add some bells or harmonica to begin a sing-a-long.

Now, compose your song!
You may even be ready for a parade!

SEW- SOW fun!

SEW- SOW fun!
"Sew" fun

Do you see what I see?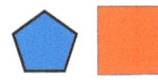

Create a yarn masterpiece by drawing a simple outline of a boat, an apple, his/her name, a heart or a football onto a piece of 12 x 12 inch burlap using a marker. Cut a piece of colored yarn about 24 inches and thread into a plastic tapestry needle. Then hand sew the outline of the object.

(If you have no drawing experience, trace a simple-lined object onto the burlap with a marker, then stitch by hand with the colored piece of yarn.)

Dripping with diamonds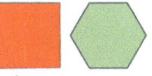

Well, maybe not diamonds but precious pasta spayed gold can be priceless!

Use tube-like pasta (ex. Rigatoni or Penne) for your necklaces.

First dye or spray paint (gold) the pasta first. When it is dry, begin stringing the pasta into the pattern of your choice. Any string works but the elastic cord allows ease of taking on and off.

Sewing felt pillows with yarn and plastic needle is a quick project. Kits can be purchased at local craft stores.

"Sow" fun

Watch 'em grow

Plants take time to grow, but in a matter of days, you can see the miracle before your eyes. Sow seed (grass, rye or bird seed) in a paper or plastic cup with soil and watch it grow in a day or so...

If it is springtime, sow the grass seeds into a cracked eggshell. Decorate the eggs shell with a face and google eyes. By the time, the grandchildren leave; they will have little grass "hairs" growing in the eggshell. Isn't this a better treat to send home rather than a duck or baby chick?

Grandma says, "Eat your worms!"

Worms to eat? Yes! Make your own edible dirt using crushed Oreos™ and chocolate pudding. Layer the pudding and crushed cookies, and top off the dirt with gummy worms to make your edible habitat.

The Magic Beanstalk

Can you climb a beanstalk? Grow one first. Hide lima beans in a wet paper towel and place in a plastic sandwich bag. Hang this in a window, as plants need water and sunlight to grow. In a few days, you will have a beanstalk.

Hide and Seek

What do worms do all day? Discover the answer when you make your own aboveground habitat. Using a plastic liter bottle cut off the top leaving a plastic cylinder to fill with dirt about 2/3 full. Layer with soil, leaves, and sand, and then top with dead leaves, before adding a little water. Find 2-3 earthworms and place in your bottle. Cover with plastic wrap, and poke a few holes in the top of the plastic wrap. The worms will mix the layers in a short time.

Yum Fun

14

Yum Fun

Road Racer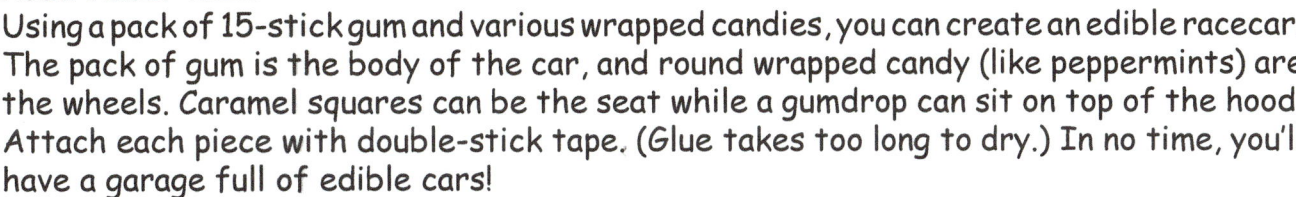

Using a pack of 15-stick gum and various wrapped candies, you can create an edible racecar. The pack of gum is the body of the car, and round wrapped candy (like peppermints) are the wheels. Caramel squares can be the seat while a gumdrop can sit on top of the hood. Attach each piece with double-stick tape. (Glue takes too long to dry.) In no time, you'll have a garage full of edible cars!

Raining outside? No problem...

Use a card table and a bed sheet to make a great **fort**. Hide from pirates as you eat lunch. Add a few flashlights and books, and spend the afternoon hidden away.

Biology for Preschool

Serve food items that resemble the body. Carrots cut in circles are like eyes; celery sticks are like bones. A cluster of grapes is like the heart, etc. Get creative, and come up with your own.

Play with your food

- **Gone fishing**

Go fishing with a pretzel stick covered in chocolate syrup or marshmallow cream (bait), and fish in a bowl of fish-shaped crackers

- **Tall Towers**

Build a tower by joining toothpicks and tiny marshmallows. Make a square using 4 toothpicks held together by marshmallows. Continue adding marshmallows at the point of intersection to the toothpicks. How tall can you make your tower?

- **Winner eats all!**

Play Tic Tac Toe using graham sticks and raisins or marshmallows.

Domino Brownies— Make a pan of brownies. Cut small rectangle shapes. Add white chocolate morsels to create the numbered dots on a domino.

- **Step back in time...**

Make your own butter by using a glass quart jar filled halfway with heavy whipping cream. Add a pinch of salt and tighten the lid. Shake, shake, and shake some more. This takes an hour or more to do but you can put it down and come back to it. It will begin to separate leaving a lump and very thin milk. Take the lump out and dry on a paper towel. Yummy Fun. Serve with crackers, and your snack is ready. Your grandchild will learn a little chemistry while enjoying a homemade snack.

This is an activity that works well if the grandchildren are coming for 4-5 days. Everyday IS a Holiday:

Choose your grandchild's most favorite holidays, and create activities and foods for each day. Plan a craft and food item for each day. See suggestions below.
This does not need to be overwhelming! It is not necessary to decorate the house when creating each event. For example, red tablecloth works for 3 of the 4 holidays listed. And if you make one box of cupcakes on the first day, then use the cupcakes for the subsequent holidays. Given are a few ideas. Also, visit your public library and get holiday videos and books to correspond to each event. (Usually no problem obtaining Halloween and Christmas books in the summer!)

Christmas

- Pull out your Christmas placemats and Christmas china

- Serve ice cream with sprinkles in crystal compote glasses

- Make presents for the parents. Ideas are pillows, weaving loom potholders or ornaments

- Decorate cupcakes with red and green sprinkles

- Watch a Christmas movie

Halloween

- Tie-dye white t-shirts in orange dye.

- Make plaster of Paris stones for the parents' gardens

- For lunch, serve "mummy dogs" (hot dogs wrapped in crescent roll pastry)

- Decorate cupcakes in orange and black sprinkles

- Read a Halloween book

- Make ghost lollipops with tootsie pops, tissue, and ribbon

Fourth of July

- Decorate flip-flops with Fourth of July jewels

- Paint the American flag with watercolors. There are 50 stars and 13 stripes. The top and bottom stripes. Are red so the flag can be seen when flying in the sky

- Fix s'mores for dessert

- Turn on the sprinkler for a little "cool" fun with water

- Light sparklers

Valentine's Day

- Mold jewelry (necklaces and bracelets) by using oven-baked clay. Clay comes in a variety of colors. Roll clay into a small balls, pierce with a toothpick. This allows the elastic cording to go through after baking.

- For lunch, cut their favorite sandwiches in a heart shapes.

- Decorate sugar cookies

- Make Valentines for each member of their family. Items to have on hand: pink paper plates, doilies, glitter, tulle, yarn, ribbons, colored art paper and jewels.

Thanksgiving

- Pinecone turkeys are fun to make. Find several open pinecones. Then at the point of the cone, glue a little red balloon for the turkey's wattle. Try to find one good protrusion to use and dot it with glue. Next, put the red water balloon over the tip of the pinecone that you glued. This will be the turkey's wattle. Pumpkin seeds make great eyes. Draw a black dot for the center of the eye. Then, put some glue on a leaf of the pinecone above the beak. Add some feathers of construction paper to finish your turkey! It's ready for the table.

- Dress for the occasion by making Indian vests (made from brown paper bags) and headbands

- Decorate the table with hand-print turkeys using finger paint for placemats

- Draw a tic-tac-toe board, and play using candy corns and other fall candies for pieces

SUM fun

You don't have to be a math teacher to introduce your grandchild to "sum" fun.

Just a fraction, please!
When you fix a PB&J, waffles , or a grilled cheese sandwich, cut into fourths, eighths or twelfths. This is a simple way to begin teaching fractions.

Keep 'em guessing.
When you serve cheerios, raisins or marshmallows, ask the child to estimate the number in this serving.....if you do this many times, you are teaching a life skill of estimating. Give the same number of items in different size glasses, bowls or jars to encourage critical thinking. If they can count to ten, have them place 10 marshmallows in a small cereal bowl, a serving bowl and then a large mixing bowl.

Easy as 1,2,3...
Use dice or number cubes and find how many "tens" you can make: 4 and 6; 3 and 7; 1+2+3+4. There are many ways to count tens.

If you have the foam number cubes, toss 3 cubes (dice) in the air and when they land ask the child to make the largest number from the digits. For example, if 1, 2 and 5 turn up, the largest number would be 521. Ask about the smallest number. This works well with small dice, too.

Domino Trains
Using dominoes, make towers and then snake lines to knock down. Teach them to play dominoes.

"When are we going to be there?" Follow me.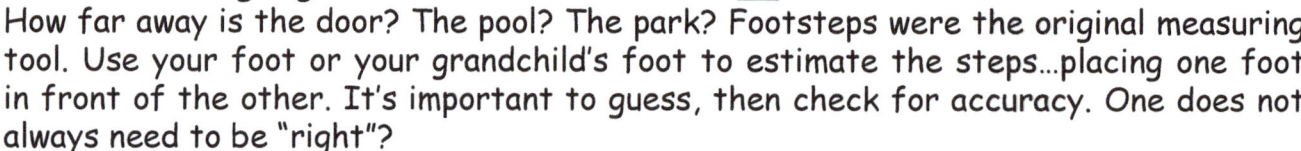
How far away is the door? The pool? The park? Footsteps were the original measuring tool. Use your foot or your grandchild's foot to estimate the steps...placing one foot in front of the other. It's important to guess, then check for accuracy. One does not always need to be "right"?
How many footsteps from the car to the back door?
How many footsteps on the frontage property?
How many footsteps from the street to the front door?
How many footsteps from the kitchen to the closest bathroom?
How many footsteps from Granny's room to their room?

Shapes, shapes, shapes...everywhere I look, I see shapes!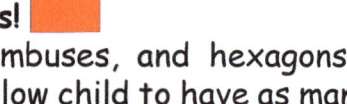
Cut out colored shapes of triangles, parallelograms, rhombuses, and hexagons for pictures. You can also cut odd shapes in different colors. Allow child to have as many as needed for their picture. Glue each shape (their new picture) onto a piece of construction paper.
Encourage your grandchild to tell you *about* their picture instead of asking, "What is it?" They often assume you should *know* what it is.

Basic Architecture

Using plastic straws and string, create different 3-dimensional shapes like a cube or rectangular prism. String can be "sucked" easily up in a straw. Careful; it scoots rather quickly down through the straw when sucking. Can you make it into a cube or rectangular prism?

Patterns...I found it!

Teach patterns in the kitchen when setting the table: fork, spoon, fork, spoon. This is an ABAB pattern. Children can look at your curtains, your ceiling, your floor and find more patterns. You can find them on your clothes and foods. Cut an orange and see the pattern of the sections. Make a fruit kabob using a pattern: apple, kiwi, strawberry, apple, kiwi, and strawberry. Look around and you will see things you've never seen before!

Making the pasta necklace, you can create your own patterns: pasta, cheerio, cheerio, pasta making an ABBA pattern.

Word Fun:

Going on a "word" hunt...

When you have finished reading the morning news, use the front page as a search for the alphabet. When the child has located each letter circle, circle with a marker. There will usually be a "q" on the first page—look hard. When they finish, connect the letters to make a figure. It usually makes a monster or ghost quite easily. Then just add ghostly eyes.

Trial by letters...

Need a quiet game for two?

Hangman is a game for two people using only paper and pencil. One, player thinks of a word and the other one tries to guess the word. The second person guesses possible letters and if the other player succeeds, then the first person writes the letters in the correct position. If the suggested letter does not belong in the word, then one element of the hangman is placed on the hangman diagram. One begins drawing the hanging man with a head, torso, left arm, right arm, then left leg and finally the right leg.

Walking and Talking Dictionary...
Vocabulary grows best when heard in context. Watch the impact you can have on your grandchild when you incorporate a few works into conversations.
For example: Camouflage-find a lizard and watch it change,

Potable-water that is good to drink

Parallel-look at leaves with parallel veins, train tracks

Dozen-find things that come in groups of twelve

Colander-to strain

Concentric circle-draw or throw rocks in a pond

Balance-equal in weight or value

Find different moments to use these words throughout the day.

Name game

Using your grandchild's name, family's name, pet's name, friend's name, create a word search. For example: Frannie, Pops, Lee, Jack, Emily, Anne Calvert, and Daniel are all hidden in the nest of letters below. Have your grandchild seek "your" names amongst random letters.

Circle all the names you can find.

AAAAAAAFRANNIEAAAAPOPSAAAAA

AAAFRANNIEAAAJACKAAAAAAAAA

AFRANNIEAAEMILYAAAAADANIELAA

ANNECALVERTAAAFRANNIEAAAAAA

AAAAAAPOPSAAAAAAALEEAAAAAA

A word search, word find, or mystery word puzzle is a word game. Letters of a word are placed in a grid, that usually has a rectangular shape. The object of the game is to find and mark all of the words hidden. The words may have been placed horizontally, vertically, diagonally or backwards. For beginners, have words written horizontally. Provide a list of the hidden words so success is achieved.

Personalized Memory Game—It's all about YOU.

Make two identical pictures of the child, their mom, dad, cousins, grandparents, etc. Glue each picture onto an index card. To play the game, place several pair of cards face down-- 4 to a row using about 3 rows. Now turn 2 cards over to see if they match. If not, turn them back over face down. If they do match, they made a pair. Now it is your turn to try to find a match. Once you find a match, the cards are yours. If the child is 4-5 years of age, try using 6-8 cards at first to experience success. Then add additional pairs to the game. This can be done with pictures of trucks, family pet, flowers or any interest your grandchild may have rather than family photos. This works well if the family is growing with a new baby coming home. Add pictures of new baby, mommy and daddy.

RE-CYCLE FUN

28

RE-CYCLE FUN

The Age of Robots….

Prior to the arrival of your grandchildren, begin saving recycled materials such as milk cartons, cereal/cracker boxes, Styrofoam meat containers, plastic lids, empty paper towel rolls, pie tins and cardboard boxes of differing sizes. With all the materials spread out, allow the children to come up with their own robot. This works well if you have a garage or basement in which they can work. Let them figure out how to make it! You will be amazed at what their robot can do.

This picture will be worth a thousand words!

And the Walls Came Tumbling Down…

Build a tower using six to eight Styrofoam meat containers of similar size and ten to twelve plastic cups. Place two cups down; add the meat tray on top, then two more plastic cups, meat tray on top. Continue layering until it falls.

How tall can you make your tower before it falls?

How fast can you build it?

Knot Fun

Knot Fun

Now I can play beauty parlor

Remember learning to braid? Now you can teach a child how to braid. Use yarn to make a thick, 20-inch long ponytail. String the yarn ponytail onto a chair to anchor it. Show the pattern of braiding using three divisions: left over middle, then right over middle and continue. You can use ribbons to work in place of yarn.

All sailors need the basics

Teach square knots and slip knots with ropes. Use a Google search or Cub Scout manual if these knots have slipped your mind.

Cat's Cradle

Search the Internet for step-by-step directions to create figures using string. All you need is two people and a piece of string tied at the end to make a loop. Once you have mastered the Cat's Cradle, try Jacob's Ladder, Cup and Saucer or Cat's Eye.

Sort of Fun

Collect the following items and see how many ways a group can be sorted. Look carefully at the markings and details to see how they fit into a group.
- Old keys-square tops vs. others; gold keys vs. silver; oval holes vs. square holes
- Pom poms-big vs. small; colored vs. white
- Lids-milk, laundry detergent, spice jar, any screw-on lids

Cut up Fun

Puzzle mania
Before you throw away that empty cereal box, here is a game. Cut up the front picture into various straight-edged shapes, and then let your grandchild re-make the puzzle. Wrapping paper with a scene works as well.

Draw lines on computer paper in the following ways for children to begin cutting.
Cut it out

- Waves

- Zigzag

- Snake

- Bumpy Road

Shapely Bag
Cut up assorted shapes using construction paper: rectangles, circles, squares and trapezoids. Children can mix and match shapes to create a new picture. Have the child glue his picture onto a piece of black construction paper.

Is it a plane? Is it a bird? No, it's a BUNNY.
Bunny-copter: Cut out the shape of a bunny's head with long floppy ears. Fold one ear forward and the other backward. Roll up the "neck" and clip with a paperclip for weight. Stand on chair and drop the bunny-copter and let it whirl.

Indians and Pilgrims coming to Thanksgiving?
Make an Indian vest: Using a brown paper sack, open it up (as if you are going to fill it), cut up the middle of one side of the sack to the bottom of the sack. Make a circle in the bottom of the sack for the neck. Cut two armholes on each side. Then you have made your vest. Now, give your child colored chalk to decorate the vest. They usually love "fringing" the ends of the sack with scissors. An Indian headband will complete the costume.

Stepping back in time...
Remember paper dolls? Show a child how to fold a paper like a fan and cut paper dolls. Then you can fold in half and cut one half of body opening the paper to see the whole body. You have just taught symmetry.

34

"So Fun for Free"

- Take a trip to your local fire department. They are always available and no appointment is needed.
- Find a body of water (lake, pond, bayou) in your home town—walk around it (if not too large) and count how many "feet" it is. Count your steps. Estimate from here to there. Throw a few pebbles, pinecones and sticks into the water. What floats and what sinks?
- Take a trip to your local library and gather an armload of books to read together.
- Take a trip to your local park or zoo and bring the picnic or the kite!
- Check out your local newspaper for free movies at the park.
- Does your city have a splash pond? If not, set up a sprinkler in the yard on those warm summer days.
- Look around for skating rinks and bowling alleys who offer free games (especially during the summer)
- Play board games with your grandchild. Although they have these apps on their parents' smartphones, it is important to teach winning and losing with your grandchild.
- Take a train ride to a nearby city (you may need a friend to pick you up) but the train is fun for all ages. This is not free but a real thrill for a child.

On the next page is a list of items to get you started on "your Granny box" to build memories with your grandchild that far outlast any toy. My hope is that you read this and thought "I could have written this book". Yes, you could have and now you can with your very own Granny box of memories.

My Frannie Box

Items to begin collecting for your Granny box:

Art paper
Butcher paper roll or brown paper roll
Colored chalk
Crayons
Colored felt squares
Dominoes
Double stick tape
Eyedropper
Feathers
Flashlight
Foam cubes
Glass beads or Mardi Gras beads
Glitter
Glue and paste
Magnifying glass
Markers
Number cubes (or dice) Dice have dots and number cubes have the digits 1-6
Oven-bake clay
Paints, brushes and watercolor paper
Piece of wool
Ping-pong balls
Scissors appropriate for their size
Straws
String
Sidewalk chalk
Things to sort: buttons, caps, pompoms, old keys
Watercolors and brushes
Wiggly eyes
Yarn

Find your interests and talents and have fun with your grandchildren.
Hopefully, you will come up with even more ideas and write your own book. Enjoy!

Frannie